AF479341

On the Job

FOSSIL TALES

Meish Goldish

CHELSEA CLUBHOUSE

An Imprint of Chelsea House Publishers

A Haights Cross Communications Company

Philadelphia

This edition first published in 2003 by Chelsea Clubhouse, a division of
Chelsea House Publishers and a subsidiary of Haights Cross Communications.

A Haights Cross Communications Company

This edition was adapted from Newbridge Discovery Links® by arrangement with Newbridge Educational Publishing.

Chelsea Clubhouse
1974 Sproul Road, Suite 400
Broomall, PA 19008-0914

The Chelsea House world wide web address is www.chelseahouse.com

Library of Congress Cataloging-in-Publication Data
Goldish, Meish.
  Fossil tales / by Meish Goldish.
      p. cm. — (On the job)
Includes index.
Summary: Introduces the science of paleontology by profiling a variety of famous fossil hunters and describing their discoveries.
  ISBN 0-7910-7411-0
1. Paleontology—Juvenile literature. 2. Paleontologists—Biography—Juvenile literature.
[1. Paleontology. 2. Paleontologists. 3. Scientists.] I. Title. II. Series: On the job (Philadelphia, Pa.)
  QE714.5 .G45 2003
  560'.92'2—dc21

                                                                                    2002015388

Newbridge Discovery Links Guided Reading Program Author: Dr. Brenda Parkes
Content Reviewer: Denny V. Diveley, Supervisor, Fossil Explainer Program, American Museum
of Natural History, New York, NY
Written by Meish Goldish

Cover Photograph: Triceratops fossilized skeleton
Table of Contents Photograph: Albertosaurus model

Photo Credits:
Cover: Francois Gohier/Photo Researchers, Inc.; Table of Contents page: Richard T. Nowitz; page 4: Francois Gohier/Photo Researchers, Inc.; page 5: Joe Tucciarone/SPL/Photo Researchers, Inc.; pages 6-7; (background) Underwood & Underwood/CORBIS, (left and right inset) The Granger Collection, New York; page 8: Bettmann/CORBIS; page 9: Paul A. Souders/CORBIS; pages 10-11: Francois Gohier/Photo Researchers, Inc.; page 13: Bettmann/CORBIS; pages 14-18: Richard T. Nowitz; page 19: Francois Gohier/Photo Researchers, Inc.; pages 20-21: Calvert Marine Museum, Solomons, Maryland; page 21: Jeffrey L. Rotman/CORBIS; page 22: Fred Bavendam/Minden Pictures; page 23: Courtesy N.C. Museum of Natural Sciences; pages 24-25: Frank Sweeney; page 27: Joe Tucciarone/Photo Researchers, Inc.; pages 28-29: Lewis Rey; page 29 and 30: O. Louis Mazzatenta/National Geographic Image Collection

Illustrations on pages 8 and 26 by Greg Harris

TABLE OF CONTENTS

WHAT'S NEW ABOUT THE OLD?

A Stegosaurus fossil found embedded in rock. Stegosauruses defended themselves with the long spines on their tails.

Fossils are the remains of animals and plants that lived in prehistoric times. We know about dinosaurs and other incredible creatures because their fossilized remains have been found and studied.

Paleontologists are scientists who study fossils to learn about life on Earth. They find fossilized bones and teeth all over the world—embedded in cliffs, buried under desert sands, and deep below the ocean floor. Each new discovery is a clue that may reveal something new about prehistoric life.

Paleontologists are always searching for more and more fossils. Sometimes, they are lucky enough to find not just parts, but an entire animal. Paleontologists have discovered woolly mammoths that were frozen in deep blocks of ice long ago. The bodies stayed perfectly preserved from prehistoric times until the present!

In this book, you will meet some interesting fossil hunters and see why their discoveries were important. You may also start to wonder: What new fossils will hunters find in the future? What will their discoveries reveal about the history and mystery of the world?

*An artist shows us what **Stegosaurus** might have looked like in prehistoric times.*

THE BONE WARS

In the 1800s, scientists were just beginning to study dinosaurs. Before that time, people had no idea what dinosaurs were. They didn't know the creatures had ever existed. In fact, the word *dinosaur* wasn't even a word in the English language!

Two leaders in the field of dinosaur study were Othniel Charles Marsh and Edward Drinker Cope. Marsh and Cope first met in Europe, when they were students. The two men started out as friends. They shared a common interest in dinosaurs and other kinds of fossils.

However, both men were also conceited, selfish, and competitive. As a result, their friendship quickly soured. How, exactly, did it happen?

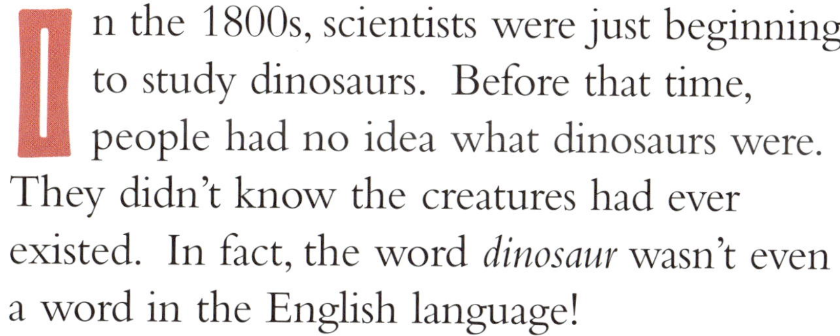

Edward Drinker Cope (inset left) and Othniel Charles Marsh (inset right). Behind them is the **Brontosaurus** *skeleton in the Peabody Museum.*

In 1870,
Marsh and several of
his students were searching
for fossils in the western United
States. They found many **specimens**,
including what looked like the wing of the
flying reptile *Pterodactylus*.

Cope also traveled west to hunt for fossils. Marsh
angrily accused him of invading his territory. The feud
grew worse after Cope hired one of Marsh's students,

Artists help us to imagine how extinct reptiles might have looked, like the pterosaurs shown in the air and the plesiosaurs shown in the water.

and then got hold of some of Marsh's fossils. The "bone wars" had begun.

Both Marsh and Cope paid work crews to find fossils to send back to them as quickly as possible. Soon the crews were feuding as well. The rival groups would spy on each other, wearing disguises to sneak into the enemy camp and steal bones.

Workers also committed **sabotage**. They sometimes dynamited each other's fossil **localities**. After a dig, they even dynamited their own localities, so the other crew couldn't dig there later.

Marsh and Cope raced each other to rebuild skeletons and name dinosaurs. Their haste led to some embarrassing errors.

In one case, Cope published a description of *Elasmosaurus*, an ancient sea reptile he had reconstructed. Marsh happily pointed out later that Cope had placed the skull on the wrong end of the animal!

Another famous goof occurred in 1877. Based on a few bones he had found, Marsh named a dinosaur *Apatosaurus*. Two years later, he named another incomplete dinosaur *Brontosaurus*. Much later, scientists discovered that both animals were actually of the same **species**.

To make matters even worse, Marsh accidentally placed the skull of a third animal— *Camarasaurus*—onto the body of his *Brontosaurus*. His mistake was not corrected for 100 years!

Two enormous dinosaur skeletons attract visitors to Dinosaur Hall at the Utah Museum of Natural History in Salt Lake City.

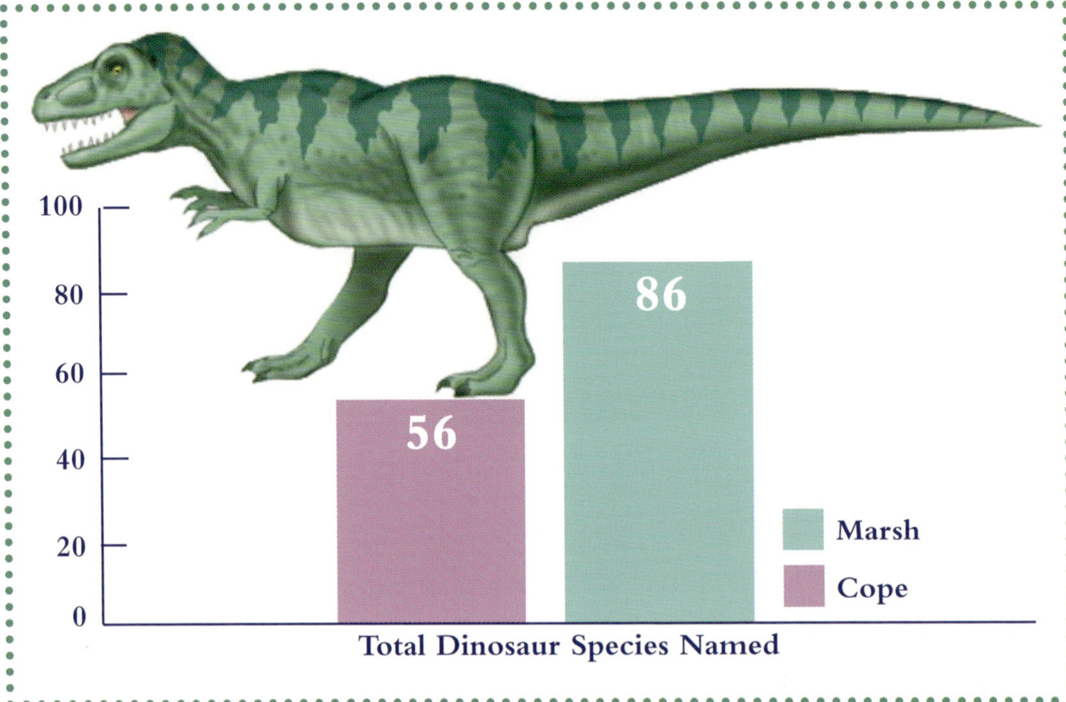

Use this graph to find out who won the bone wars.

In 1890, a New York newspaper published a series of articles about the "bone wars." Marsh and Cope insulted each other in print. Cope called Marsh's work a "remarkable collection of errors and ignorance." Marsh called Cope "my bitter enemy" and said he had "doubts as to his sanity."

Within 10 years, both men had died. Despite their feud, they actually helped the field of paleontology by identifying many new species of dinosaurs. Today, museums are filled with the bones that these scientists discovered.

SEARCH FOR A LOST LAND

Barnum Brown was known as a man who could "sniff out fossils." He attended Columbia University in New York City but dropped out to hunt fossils. Later he became an assistant paleontologist at the American Museum of Natural History in New York City.

Brown traveled widely in search of fossils. He made some amazing discoveries during his career. In 1902, he found the bones of an unknown dinosaur in Montana. The dinosaur was given its new name: *Tyrannosaurus rex*.

Barnum Brown shows off the rib of a dinosaur fossil he discovered in Wyoming.

Dr. Philip Currie and Eva Koppelhus sift through rocks in search of dinosaur skeletons.

In 1910, Brown made another remarkable discovery. In the **badlands** of Alberta, Canada, he came across a large number of dinosaur fossils. They were skeletons of *Albertosaurus*, a creature named for the place where the bones were found.

Brown was very excited. It was quite rare to find so many dinosaur skeletons in one spot. He packed up all the *Albertosaurus* fossils and sent them to the American Museum of Natural History.

Brown died in 1963. A mystery about the dinosaur bones arose after his death. No one knew exactly where in Alberta they had been found. Brown had never drawn a map or recorded the spot where they'd been discovered.

In the 1990s, two scientists—Dr. Philip Currie and Eva Koppelhus—set out to find the place where the *Albertosaurus* bones had been unearthed. They hoped that more dinosaur fossils might still be hidden there.

Fossil hunters used old photographs to locate a site on the Red Deer River in Alberta, Canada.

Paleontologists carefully record the location of every fossil bone they find, using a grid frame to map the pieces.

Currie and Koppelhus began by searching the American Museum of Natural History for clues. They eventually found four of Brown's old photographs of his work site. They also found notes he had taken in the field.

In 1997, Currie and Koppelhus organized a team of scientists. They rafted down the Red Deer River in Alberta. They tried to match the scenery they saw with details in Brown's photographs.

For days, the search turned up nothing. Then, one very hot day, Currie saw a hill that resembled a hill in Brown's photos. When he went back and told the team, they left their rafts and roamed the land for hours in the hot sun. Finally, on a mountain ridge, they discovered a pile of dinosaur bones!

Immediately, the Currie and Koppelhus team began to dig for more bones. They found plenty. By the year 2001, parts of 12 different *Albertosaurus* skeletons had been dug up.

Currie and Koppelhus's discovery revealed quite a lot of information. The bones were found together, indicating that these dinosaurs probably lived and hunted in packs. Also, their teeth revealed that *Albertosauruses*

Albertosaurus *is named for the place the fossil was first found, in the badlands of Alberta, Canada.*

could bite better than they could chew. They probably
swallowed large chunks of animals dead or alive.

Currie and Koppelhus are puzzled as to why so
many of the dinosaurs died in one place at the same
time. They hope that as they continue to dig at this
site, they will find more fossils that provide clues to
the answer.

NOTHING BUT THE TOOTH

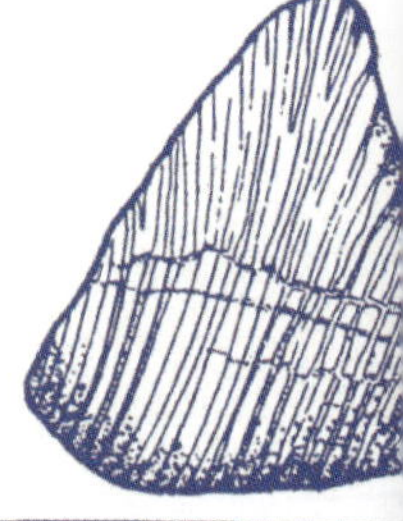

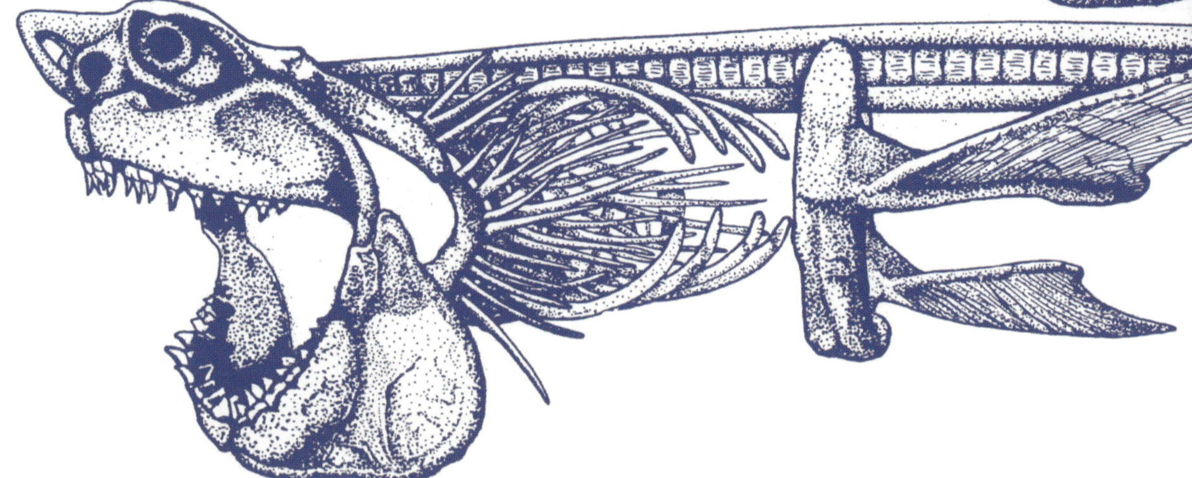

One day in 1992, an amateur fossil hunter named Rebecca Hyne was working alone on a huge mound of rocks and sand on a beach in North Carolina. Suddenly she noticed the outline of a jagged-edged shark's tooth. It was 5 inches (12.7 centimeters) long.

For a week, Hyne dug further into the mound. She found 25 giant teeth in all. She knew that they came from a megalodon shark, the ancient ancestor of the great white shark. *Carcharodon megalodon* was twice the size of today's great white. It was more than 40 feet (12.2 meters) long. Its jaw, when fully opened, 6 feet (1.8 meters) wide!

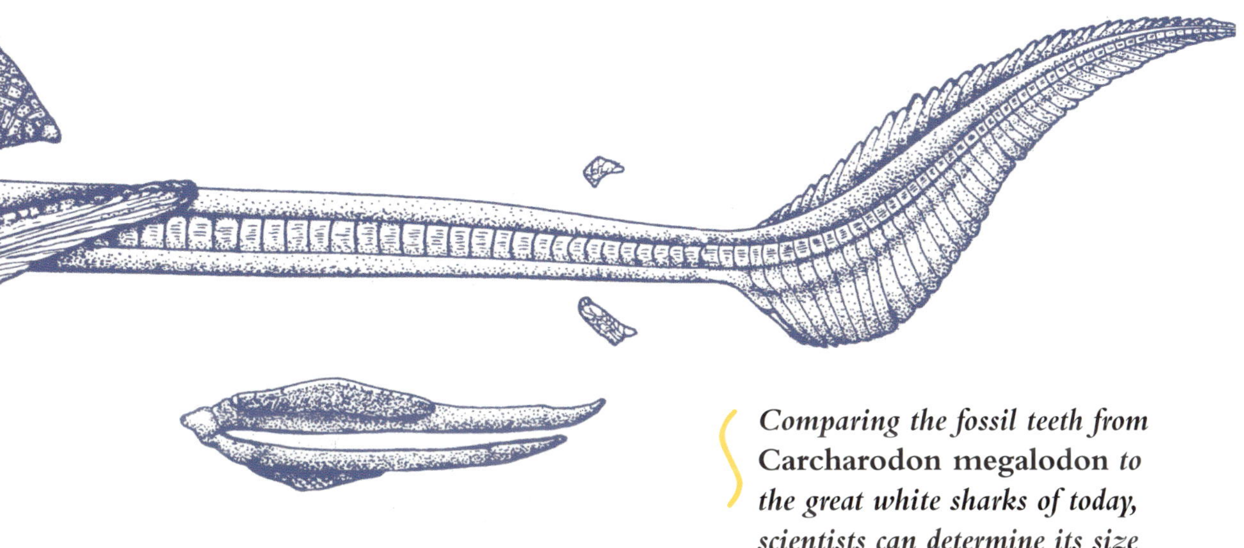

Comparing the fossil teeth from Carcharodon megalodon *to the great white sharks of today, scientists can determine its size and probable length.*

Shark teeth are very precious to paleontologists. Since sharks have **cartilage** instead of bones, it is difficult to find fossils of their skeletons. The best way scientists can learn about ancient sharks is through fossils of their teeth.

Scientists believe the megalodon used its sharp teeth mainly to eat large whales. A megalodon could easily swallow any other kind of shark. Megalodon

The great white shark is believed to be a modern relative of the ancient megalodon.

teeth are so big, the first ones discovered hundreds of years ago were thought to be serpents' tongues.

Hyne's megalodon teeth were the best set ever found. Many private collectors offered to buy the teeth. Hyne, however, believed they belonged in the North Carolina Museum of Natural Sciences.

Today, the museum proudly displays its collection of megalodon teeth. They are mounted in a cast of the shark's jaw.

Hyne has collected over 20,000 fossils during her weekend hunts. She describes the megalodon teeth as "the most outstanding thing I've ever found. It's always a thrill to see a tooth lying there—something that never had human hands on it for millions of years."

Rebecca Hyne (left) and Betsy Bennet, director of the North Carolina Museum of Natural Sciences, examine Hyne's megalodon teeth.

BUILDING AN EARTH-SHAKER

What is the biggest dinosaur that ever lived? For a long time, paleontologists believed it was the *Seismosaurus*. This prehistoric creature bears the nickname "Earth-Shaker." That's because when it walked, it's thought that the earth shook.

Bone fossils of the *Seismosaurus* were first discovered near San Ysidro, New Mexico, in 1979. They were shown to a paleontologist named Dr. David Gillette. He ordered a complete **excavation** of the site. Diggers found more *Seismosaurus* fossils: 25 vertebrae, part of a pelvis, five **chevrons**, a shaft fragment from a large bone, and many ribs.

First, workers cleaned each fossil carefully to remove any particles, like sand or rock, that were embedded in the bone. They used tools, such as chisels and needles. Then the bones were shipped to Canada, where a model of the dinosaur was made.

The life-size model of Seismosaurus is constructed on adjustable poles, so that it can be posed or wrapped around to fit into a shorter building.

The fossils showed that the *Seismosaurus* stretched about 130 feet (40 meters) from head to tail. It weighed 65 tons (59 metric tons). No wonder the earth might shake when it walked!

In July 1999, a company called Dinosauria International built a life-size model of *Seismosaurus*. It was the largest dinosaur model ever constructed. The bones were made of Styrofoam. Plaster bones would have made the model collapse under its own weight.

The *Seismosaurus* model was much too big to be constructed inside a building, so it was built outdoors. Workers first erected poles to support the dinosaur's long frame. Then they constructed the animal's neck, tail, ribs, and legs. Later, they added the dinosaur's feet and head. The *Seismosaurus* is still the biggest, but not the longest dinosaur ever. In 2000, paleontologists in Argentina dug up the bones of a plant-eating dinosaur thought to measure over 154 feet (47 meters) from head to tail!

SIZING UP DINOSAURS

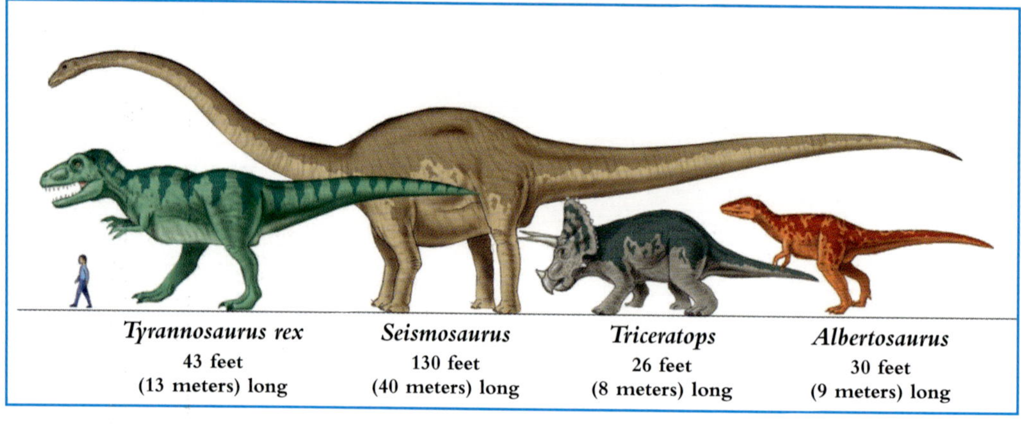

PICTURING DINOSAURS

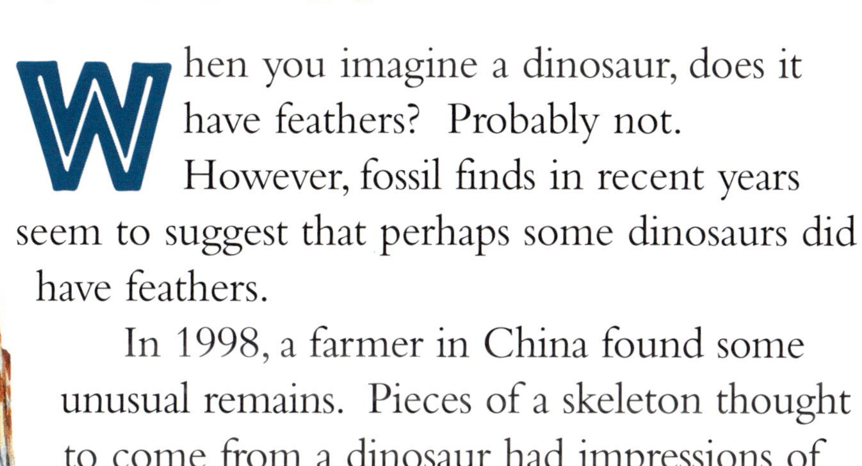

When you imagine a dinosaur, does it have feathers? Probably not. However, fossil finds in recent years seem to suggest that perhaps some dinosaurs did have feathers.

In 1998, a farmer in China found some unusual remains. Pieces of a skeleton thought to come from a dinosaur had impressions of featherlike **filaments**, or threads, pressed into the bone. The impressions were around the dinosaur's legs, arms, shoulders, and back.

*An artist's view of a small, birdlike dinosaur with feathers called **Caudipteryx**. It could not fly, but it ran on its strong hind legs.*

Scientists thought that the new species, called *Beipiaosaurus inexpectus*, was the largest known type of theropod, or feathered dinosaur. Scientists have uncovered other theropods in the same area. But unlike them, this one is thought to have been a plant-eater.

The discovery added fuel to an age-old debate: Did birds come from dinosaurs? Xing Xu, a leading Chinese paleontologist, thinks the answer is yes.

"It suggests lots of dinosaurs have feathers, and supports the **hypothesis** that the bird feather is derived from such kind of filamentary structures," Xu has said.

Not all paleontologists agree, however. Larry Martin of the University of Kansas has a different theory. He thinks the filament impressions are not feathers at all. Rather, they're a sign of muscle threads under the dinosaur's skin.

The imprint in this rock could be of a **Caudipteryx.**

Yet another theory is that young dinosaurs had feathers to keep warm. Then they shed their feathers as adults.

Perhaps fossil discoveries in the future will settle this debate and others. Each time scientists find new remains from ancient times, they gain new clues and ideas about the mysteries of prehistoric life.

Who knows? Maybe one day you will stumble across a fossil that will change the way the world thinks about the past!

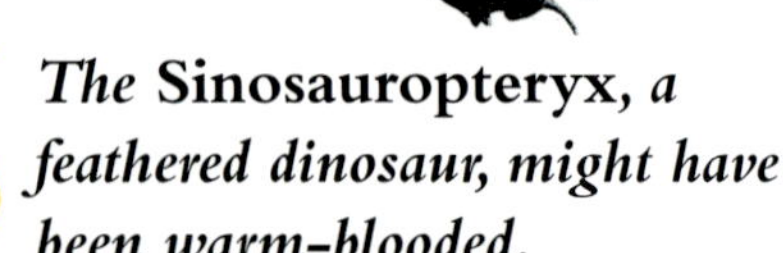

The Sinosauropteryx, a feathered dinosaur, might have been warm-blooded.

GLOSSARY

badlands: barren regions in the western United States and Alberta, Canada, where the dry soil and soft rocks take on strange shapes

cartilage: a tough tissue forming part of the skeleton

chevrons: V-shaped bones that are near an animal's tail

excavation: a digging up of the earth

filaments: very slender threads or threadlike parts

fossils: hardened remains or traces of plant or animal life from long ago, preserved in the earth

hypothesis: an assumption about something unknown made in order to test whether it is true

localities: places where fossils have been found

paleontologists: people who study fossils to learn about prehistoric life

sabotage: the purposeful destruction of machines or work materials, as during a dispute

species: a distinct kind, variety, or type of living organism

specimen: a part of a whole, used as an example

INDEX